A Science Museum
illustrated booklet

EARLY MACHINE TOOLS

by K. R. Gilbert
M.A., D.I.C.

Her Majesty's Stationery
Office London 1975

I SBN 0 11 290182 4

Cover: Maudslay bench lathe (*see plate 15*)

Note: All the items illustrated in this booklet are part of the Science Museum Collections

Introduction

A machine tool is a machine designed to cut or shape metal or other substance, and by its action replaces the skill and strength which would be necessary to carry out the same operation manually with the aid of a hand tool.

Skill applied in the original design and construction is reproduced in the work done by the machine and the application of mechanical means of controlling the position of the cutter in relation to the workpiece makes it possible to fashion objects without the skill which a craftsman would need to carry out the same task by carving or filing. Indeed such machines as the Bersham boring mill (10), which was built in 1775, could do work with an accuracy then unattainable by any other means.

The application of power provided by the horse, the water wheel, and later by the steam engine made it possible to machine objects of great size beyond the capacity of a man using his own muscle power.

The high speed of working with power-driven machine tools and the utilisation of skill built into the machine made possible manufacturing processes which would not otherwise have been economically practicable, as exemplified by Bramah's machine tools (11) specially designed for manufacturing his patent lock. It was claimed that Brunel's block-making machinery (13) operated by 10 men did the work of 110 skilled blockmakers.

This booklet illustrating exhibits in the Machine Tool Collection of the Science Museum traces the development of machine tools from quite primitive lathes to machines which begin to have a modern look, were rigidly constructed of metal, and which were indispensable for the construction of machinery. Without them the manufacture of the machinery, which was the technological basis of the Industrial Revolution, would have been impossible.

The date chosen to end the period covered by this booklet is 1831, when Henry Maudslay died. He was the leading machine builder of his time. His principles of machine construction, namely: absolute rigidity, the use of true plane surfaces and accurate screw threads (14), and workshop techniques such as the practice of precise measurement ensured the successful production of accurate and reliable machinery, beginning with Marc Brunel's mass production plant for pulley blocks, and passed into engineering practice through the influence of the many distinguished engineers, among them Whitworth and Nasmyth, who had been his employees. One of them, Richard Roberts, was already working independently in the period, and his machines (16, 17, 18) have a functional appearance, dispensing with the ornamentation partly derived from classical architecture, which was a feature of Maudslay's machines (15). The most ancient machine tool is the lathe, which must have been used from at least c. 700 B.C., the date of an Etruscan wooden bowl, which was made by turning. The earliest known representation of a lathe is a relief carving in a late Egyptian tomb of the 3rd Century B.C. showing a wooden furniture leg being turned between centres. The workpiece is rotated by a cord which passes round it and is pulled to and fro by the turner's assistant, as in the primitive Chinese lathe (1), which was still in use in modern times. The next four illustrations show how the application of motive power to the lathe evolved.

The invention of the bow drive (2) enabled the turner to work without an assistant, then the use of the spring pole (3) freed the hand used to work the bow, so that he could use both hands to control the tool; and in the great wheel lathe (4) the power was more effectively supplied by an assistant, eventually to be replaced by an electric motor.

In the first three methods the direction of rotation of the workpiece alternated so that cutting only took place for half the time. Recent research on certain metal objects of Roman date shows, however, that they were probably turned in a lathe with continuous rotation, but no such lathe or representation of one has so far come to light. Continuous rotation of the workpiece, which made the use of a mechanical tool holder practicable, could also be produced by the turner himself by means of a treadle acting on a crankshaft with flywheel (5), a development which had taken place by 1500.

The ornamental turning lathe (5) shows the early employment of mechanical control obtained by the use of the slide-rest and the copying of special forms by the use of rosettes. The copying principle is later seen in the medallion lathe (8) and the engraving machine (12). The fusee engine (7) employs the principle of the lead-screw for copying screw threads and with the wheel-cutting machine (6) represents the contribution made by the clockmaker to the development of machine tools, for the clock was the earliest machine of any complexity assembled from components of metal, which had to be cut to shape.

Three other machine tools, which were driven by water power, were introduced in the Middle Ages: the reciprocating saw, the trip hammer, and the boring machine. A machine for boring out logs to make water pipes was illustrated in c. 1430 and very substantial machines for boring out solid castings to make cannon (9) were used early in the 18th century and introduced into England in 1770. The original method of boring guns was simply to use the boring machine to clean up the hole in a barrel cast hollow over a core, and this procedure was applied in the early 18th century to boring out cylinders for Newcomen steam engines. As the boring tool was supported at one end only, it inevitably tended to follow the hole rather than straighten it; and its weight made it cut mainly on the lower part of the cylinder, so that the surface was not truly circular in cross section. Cylinders bored with such machines were adequate for Newcomen engines, but it was not possible to achieve the accurate fitting of piston in cylinder required by James Watt for his much more efficient engine.

The solution of the difficulty lay in using a boring bar supported at both ends and drawing the cutter along it. The cutter head thus supported by a straight and rigid bar was able to bore out a true cylinder. This machine (10), invented by John Wilkinson, was used for boring the cylinders from 1775 until 1800, when Watt's patent expired. It was of crucial importance in the Industrial Revolution, as it made possible the construction of Watt's rotative engine and hence the utilisation of steam power for driving machinery.

The persistence of the beam type of steam engine has been explained by its having few flat surfaces, which were costly to produce by hand methods. The invention of the metal-planing machine offered the possibility of making slide-valve surfaces and of constructing machinery with parts sliding in guide-ways. The invention of the planing machine in the first quarter of the 19th century has been attributed to Matthew Murray, James Fox, Joseph Clement, and Richard Roberts; but Roberts's machine (17) is the earliest now surviving. His self-acting slotting machine (18) derives from Brunel's mortising machine and arose from one of the requirements of the emerging railway technology.

Early in the 19th century Maudslay, Murray, and Fox were manufacturing machine tools not only for their own use, but for sale to other engineering firms and for use in mints, arsenals and dockyards. Machine tool building had become an industry.

1 Chinese Wood-turning Lathe

This model represents a lathe seen in Hankow in 1909. With the exception of the iron centres it is made entirely of wood and consists of a long rectangular frame on trestles. One centre is embedded in one of the end members and the other is in a cross-piece which is wedged in grooves in the sides of the frame and can be moved to accommodate workpieces of different length. The tool-rest is a bar which is supported at one end on the cross-piece and at the other on a movable bridge.

The work is rotated by an assistant who pulls alternately on the ends of a cord passing a few times round the work, but the turner can only take a cut when the work rotates towards him.

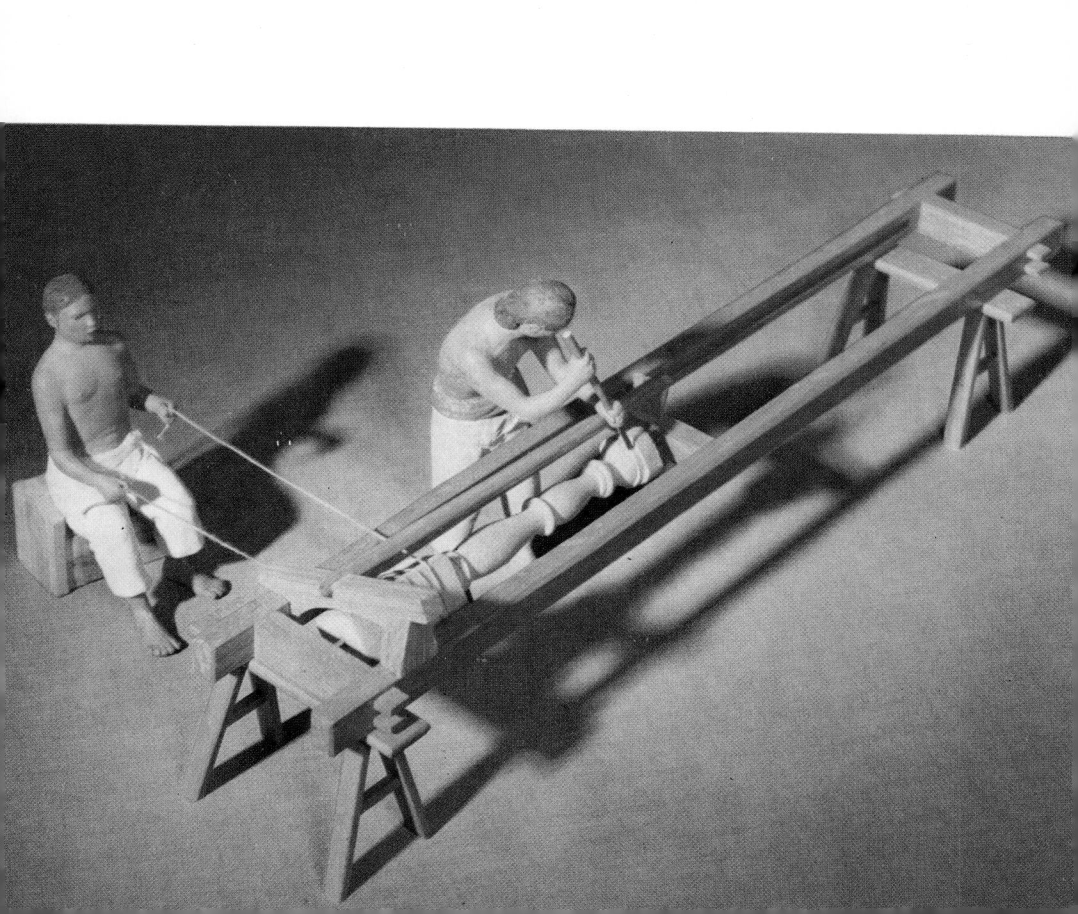

2 Primitive Lathe from Egypt

This is an example of the primitive type of bow lathe which has been used in the East since ancient times up to the present day.

At the left hand end of the wooden base board is fixed a transverse wood block which is connected with a similar but movable block at the other end by means of a stretcher. The iron centres are carried in these blocks and the distance between them is adjusted to suit the length of the workpiece by moving the loose block towards or away from the other and then fixing its position by wedging the stretcher in the square hole through which it passes. A long iron bar is laid across the blocks parallel with the centres and this serves as a tool rest.

The turner sits before the lathe on a slightly raised seat with his heels on the stretcher and his toes on the rest bar. He holds the handle of the tool in his left hand and guides the cutting end by the toes of one foot. The work is rotated by means of the forward and backward movement of a bow held in his right hand, the string of the bow passing around the work so that the direction of rotation is periodically reversed. Cutting is performed only when the work is turning towards him.

3 Pole Lathe

Pole lathes came into use not later than the thirteenth century and may have been the European development of the primitive Eastern lathe. In the Eastern lathe the turner is seated on the ground, but in the pole lathe the centres have been raised to suit the standing posture and the hand-operated bow has been replaced by a spring beam, or pole, and treadle. The lathe bed consists of two oak beams mounted on posts and the headstocks are massive blocks of oak passing between the beams and each carrying an iron centre. One headstock is fixed; the other is movable and may be clamped in any desired position. A simple hand-tool rest is provided. The driving cord is attached to the free end of the pole, passes around the mandrel, or the workpiece if a mandrel is not in use, and then through the bed to the pedal below. The rotation of the work is thus alternately towards and away from the operator who cuts only when it is rotating towards him. This lathe was made about 1800.

4 Great Wheel Lathe

This lathe forms part of the equipment exhibited in the reconstruction of a wheelwright's shop. It came from Barley, Herts. where it was used until 1949 for turning the elm hubs of waggon wheels. The lathe was driven by an assistant who turned a crank handle attached to the flywheel mounted in a stand 8 ft away from the lathe. The pulley bolted to the flywheel drives by means of a rope a pulley on the lathe mandrel.

Such a lathe is illustrated for the first time in a book of trades, with descriptions in verse by the poet and mastersinger Hans Sachs, which was published in 1568. Lathes driven by horse power and by water power were recorded in Nuremberg in the 16th century.

5 Ornamental Turning Lathe

This treadle lathe was probably made in the late seventeenth century. The five-speed pulley on the mandrel may be driven rapidly for plain turning by means of the treadle and flywheel, which is made from a single piece of timber. The slide-rest has a single slide adjustable by means of a screw.

The nose end of the mandrel runs in a cylindrical collar carried by a lever pivoted below the bench, and the pointed back end is supported in a hollowed adjustable centre. When rose-turning the collar is released by removing a locking screw so that it is free to swing about the pivot. The mandrel carries a group of ten rosettes between the pulley and the headstock, any one of which may be made to press against an iron rubber, which may be fixed anywhere along a bar running parallel with the mandrel. The rosette is kept pressed against the rubber by means of a spring so that a stationary tool applied to the work cuts a wavy line. For rose-turning the mandrel is rotated slowly by means of the handle attached to the pulley at the front of the lathe bed.

Five guide-screws of different pitch are cut on the rear portion of the mandrel to the left of the pulley. Below each of them is a wooden bearing which has the corresponding thread impressed upon it, so that the one selected may be used as a screw key. If one of these keys is raised, by inserting a wedge below it, to support the mandrel, the latter advances as it rotates, so that a cutting tool held against the work mounted in the chuck will cut a screw. This is the oldest method of cutting screws in the lathe.

Machines such as this with ingenious chucks for oval turning and the like were made primarily for amateur turners of means rather than for the artisan.

6 Wheel-cutting Machine

The earliest method of manufacturing clock wheels was to mark off the blank by means of a radial ruler, pivoted at the centre of a plate divided concentrically in numbers of teeth in common use; the teeth were then filed out by hand. In 1672 Robert Hooke mounted the divided plate so that it could be rotated, added an index, and applied a revolving file or milling cutter to the edge of the blank.

In this machine the dividing wheel is inscribed with 25 concentric circles divided for cutting wheels with from 31 to 78 teeth with some exceptions. The underside of the plate is inscribed with the name of the maker, Humphrey Marsh de Highworth, and with an unfinished Gunter's quadrant (an atronomical instrument) and a perpetual calendar. From the leap years given it is evident that the calendar was begun between the years 1668 and 1672. The work on the quadrant was however abandoned and the brass plate used to make the present machine presumably about the time of Hooke's machine.

The circular plate is attached to a spindle, which carries the blank. The spindle passes through a T plate, the ends of which are slotted and secured by lock nuts to the iron frame and which can be adjusted by a screw to take blanks of different sizes. The cutter is mounted in a frame on bearings so that it can be swung into the blank as the cutting proceeds. The cutter is driven by the handle through gearing.

This is the earliest known gear-cutting machine, but there is a reference as early as 1540 to a machine for cutting clock wheels.

7 Fusee Engine

This is a small screw-cutting lathe for clockmaker's use, dating from c. 1775, which was designed for cutting fusees. A fusee is a tapering drum of hyperbolic section with a screw-like groove cut on it, on which is wound a cord connecting the main spring of a clock or watch with the driven mechanism. It was formerly used for maintaining a constant driving torque as the spring unwinds.
The engine is intended to be mounted in a vice, and when the handle is turned, it drives through pinions the chuck holding the fusee blank. At the same time a threaded block travels along the screw cut on the shaft. The block is attached to a lever, which is pivoted at the front of the frame, and as it moves across, it carries with it the table on which the cutting tool is mounted. The point of attachment of the lever to the table can be adjusted, so that the pitch of the thread cut on the fusee can be varied. The screw in the fusee engine can thus be made to generate any required thread on the fusee blank. The tool holder is free to slide, so that the cutter can be kept in contact with the blank.

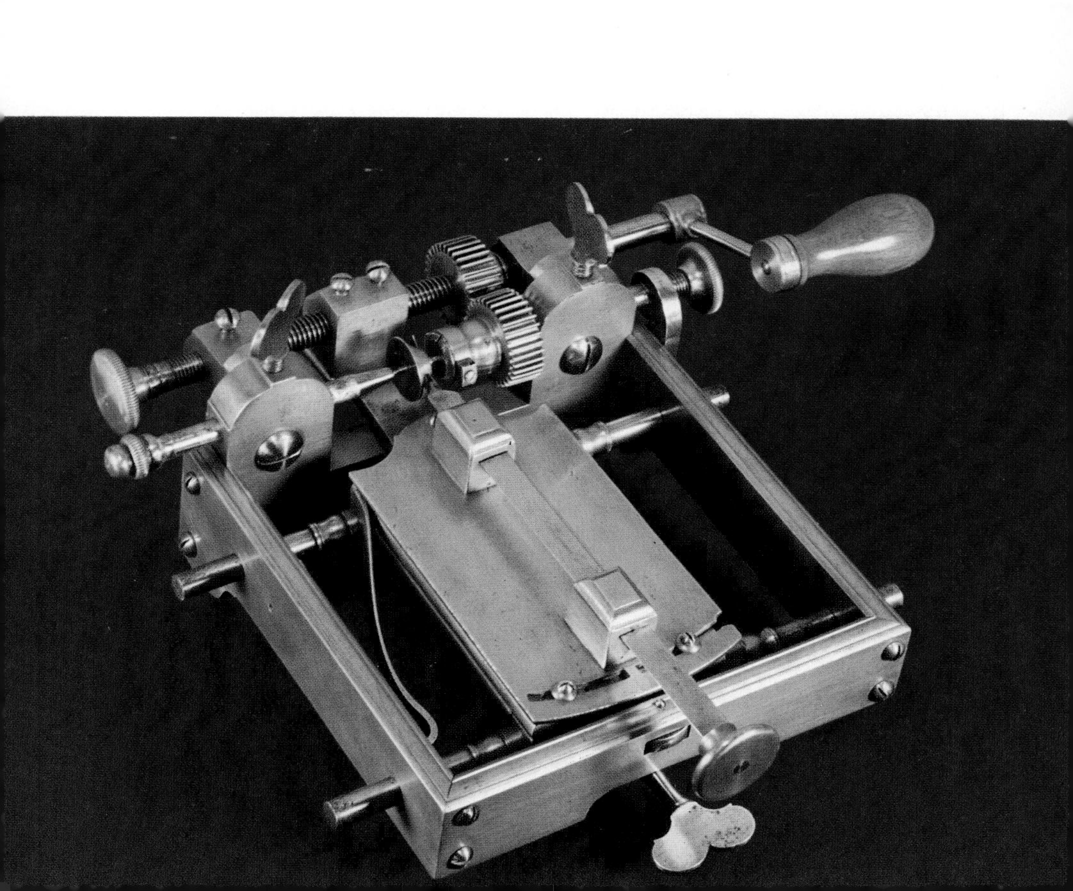

8 Medallion Lathe

This French medallion lathe was built in about 1760. The matrix is held in a chuck on the rear end of the headstock spindle on the left and the medallion blank is mounted similarly on the front end. The spindle is able to move longitudinally, but under the influence of a spring is pressed backwards so that the matrix is held in contact with a copying point. Both the copying point and the cutting tool holders are mounted on slides restrained by springs and are caused to move slowly out from the axis of the spindle by flexible wires attached to grooved pulleys. These are rotated by means of a double worm and worm-wheel train enclosed in an oil-retaining case and driven by a round belt from the spindle pulley. Both the copying stylus and the tool in effect trace spirals on the matrix and the blank.

The lathe spindle moves forwards or backwards as it rotates so that the tool cuts contours on the blank corresponding to those being traced out by the copying point on the surface of the matrix. The high points become low points and vice versa. The gearing is such that the medallion being cut is reduced in size compared with the matrix. The machine can also be used as a rose engine and for screw cutting in a similar manner to No. 5, using the second tool rest, which is seen on the right of the photograph.

9 Gun Boring Machine

This model gun boring machine is a copy of a model in the Royal Artillery Institution, made in 1782 but destroyed in World War 2, and represents the early method of boring guns at the Royal Arsenal, Woolwich.

The gun is missing, but would be rotated by the shaft protruding from the wall on the left, the muzzle end being supported by a ring guide on the left-hand end of the bench. On the other side of the wall is a horse mill which provides the driving power.

A 'D' bit probably used for finishing is shown in place and is fed into the bore by means of a rack and pinion operated by the hand wheel, the direction of the bit being kept as accurate as possible by the use of steel guides mounted on the heavy timber bench. A simple form of slide-rest providing longitudinal movement for the tool is used for outside machining of the guns when necessary. This is situated alongside the boring bit.

Henry Maudslay, who was employed at the Woolwich Arsenal for 5 years until 1789, was presumably familiar with the gun-boring machine and its slide-rest, which he employed so effectively later.

10 Cylinder Boring Mill

This model based on an original drawing represents the boring mill established by John Wilkinson in 1775 at Bersham near Wrexham, the first mill in which a boring bar supported at both ends was used. Its invention had an important influence upon the progress of contemporary engineering practice for it enabled cylinders to be bored with much greater accuracy than had hitherto been possible and all Watt's early steam engine cylinders were bored upon it. The water wheel drove two boring machines and two lathes and the larger boring machine with a sectioned cylinder is on the side of the wall shown in the illustration. Next to the boring machine is a facing lathe.

The large driving wheel is keyed to the squared end of the boring bar which is supported in a bearing at each end. Mounted on the bar and rotating with it but free to slide along is the cutter-head. This is fed forward through the cylinder by means of a rod passing down the centre of the hollow bar and attached to the head by means of a pin which passes through the longitudinal slot. The other end of this rod is prevented from rotating by a cross bar resting on two wooden beams and is attached to a rack. Rate of feed is governed by a weighted lever acting through a pinion engaging with the rack. The cylinders are supported in heavy timber cradles and are secured with chains.

The 15 ft long boring bar is the only part of the machine to have been preserved and is illustrated below. It has however at some time been modified for screw feed actuated by gearing.

11 Sawing Machine

In 1784 Joseph Bramah patented a very successful lock, in which the key depresses five pairs of spring-loaded steel strips which slide in slots in a brass barrel. The strips are also slotted and arranged to prevent the rotation of the barrel, and hence the bolt from moving, unless they are depressed to the correct positions.

The complicated mechanism could only be satisfactorily and economically manufactured by a well designed series of machine tools. To this end, Bramah, with the assistance of Henry Maudslay, set up one of the earliest factories for precision metal working.

This machine was made in the works of Joseph Bramah in c. 1790 for the purpose of cutting slots in the barrel. The saw is held in a metal frame that travels between adjustable V-slides and is reciprocated by the double-handed lever by man power. The barrel to be slotted is placed in the bush holder and this is then placed in the main holder indexed to cut 4, 5, 6, 7, 8, 10 or 12 slots in a barrel 1¼ in diameter and from 2 in to 8 in long. The holder is raised and lowered by a long lever at the side of the machine in order to bring the lock barrel into contact with the saw. The latter can be seen stretched between clamps at each end of the frame. Its thrust is resisted by a steady mounted at the mid-point of a lever pivoted on one side of the machine and held at the other by a clamp with a handle. This is released at the completion of the last cut, so that the lever can be moved away, the saw blade taken out, and the barrel removed from the machine.

12 Engraving Machine

This machine, for preparing the steel master punches from which the dies for coining are made, was purchased in 1824 for the Royal Mint. It was made in Paris and is an example of the 'tour à portrait' invented by Hulot about 1800.

The machine is in principle a copying lathe, in which the original and the copy are mounted in the same plane, on the ends of parallel horizontal axes which are coupled together and rotated at the same speed. A bar, pivoted at one end on a universal joint, passes in front of these and has mounted upon it a tracing point and a cutting tool, which are set opposite to the centres of the original and the copy respectively. While the medallion and copy are revolving the tool bar is gradually lowered by a screw at the end of a train of gears so that the tracing point traverses the surface of the original from the centre to the circumference, moving inwards and outwards as it encounters the various features of it. The cutting tool also traverses the face of the copy, in a similar manner, but the distances moved through, both transversely and horizontally, are reduced in the ratio of the distance of the tool and tracer from the pivot of the tool bar, thus producing a reduced copy of the original as nearly like it as the necessary size and shape of the tracer and tool will permit.

The medallion used as the original, which is made as large as possible, would be a reproduction of the artist's design, made in a material sufficiently hard to withstand the rubbing of the tracing point. The copy when taken from the machine requires touching up by hand in order to finish those parts which the machine is not able to reproduce exactly.

13 Mortising Machine

This machine, which was built in 1803, is one of the group of machine tools known as the 'Portsmouth Block-making Machinery'. These machines were designed for the mass production of pulley blocks for use in naval vessels. They were designed by Brunel and built by Maudslay.

This machine chisels slots for the sheaves in two wooden blocks starting from the larger of the two holes drilled by the boring machine, which preceded it in the manufacturing process. The other hole was for the sheave pin. The mortising machine can also be used for making blocks with two sheaves. It was one of the first machines to be fitted with a cone clutch and brake. By its means the shaft may be disconnected from the flywheel and the machine quickly brought to rest when it is necessary to change the blocks. Rotation of the shaft imparts a reciprocating motion to the chisels, which carry projecting tongues to clear the chips from the mortises. The machine is capable of working at more than 400 cuts per minute.

The blocks are held in a carriage which can slide horizontally by the action of a lead-screw. A nut on the lead-screw is bolted to a ratchet wheel which is advanced by the movement of a toothed lever actuated by a cam on the main shaft. The blocks thus advance after each cut of the chisels.

When the carriage has moved a predetermined distance, a weighted lever, which it has been supporting, falls and lifts the toothed lever from the ratchet wheel, thus finishing the cut. The mortising machine was one of the first self-acting machine tools and was also the largest machine tool for its time—the flywheel is 6½ ft in diameter—to be constructed entirely of metal.

14 Screw-cutting Lathe and Screw Originating Machine

This lathe was constructed at the end of the eighteenth century, and is believed to be the first workshop machine in which Henry Maudslay combined a lead-screw and change wheels for reproducing screw threads.

The bed consists of two triangular bars, which carry the headstock, tailstock, and slide-rest. The lead-screw is situated between the bars and was presumably geared to the lathe spindle by change wheels (now missing).

The small appliance shown in front was made by Maudslay in about 1800 for originating screw threads. A cylinder of hardwood or soft metal is pushed into a metal tube and turned so that it engages with a crescent shaped knife which projects into the tube and is set at the appropriate angle with the axis. The knife in cutting into the cylinder causes it to traverse and a screw is thus generated which is deepened and cut to the required form by a following chaser. Copies in steel are then made of the screws thus produced.

The knife is carried by a holder secured to a large disc provided with a tangent screw adjustment finely graduated so that the knife can be set at the exact inclination. The depth of penetration of the knife and of the chaser are also set by screws.

15 Bench Lathe

This treadle lathe with triangular bar bed is of a type manufactured by Henry Maudslay for sale to amateur mechanics and small workshops. The name of the company 'H Maudslay & Co' engraved on the brass plate on the headstock indicates that this lathe was sold between 1812 and 1820.

The bed is carried on two short standards fixed to the bench and, except when a gap is required for work of large diameter, slides into a triangular hole in the body of the headstock. For ordinary purposes the slide-rest fits directly upon the bed, but for screw-cutting it is mounted, as shown, upon an adaptor of triangular section which protrudes at right angles to the bed to which it is clamped. In this position the long lower slide of the rest is parallel to the line of centres and the traversing screw is connected through a keyed shaft to an appropriate train of wheels which is mounted upon a radial arm and driven from a pinion on the mandrel. The upper slide of the slide-rest is carried on a swivel and can be set at any desired angle.

The lathe is mounted on a bench with drawers and is equipped with a two-jaw universal chuck, and various other chucks, tools and toolholders.

16 Slide Lathe

One of the features of this lathe built by Richard Roberts in 1817 is the self-acting saddle which slides directly upon the lathe bed, and is thus able to traverse its whole length. The ordinary slide-rest of the period was a self-contained appliance, clamped to the bed where required, and only able, without being reset, to turn work of the length commanded by its short screw.

The saddle is traversed by a long screw carried in bearings at the front of the bed, and driven from the mandrel through a variable speed gear and a bevel wheel reversing and disengaging gear. It is fitted with a long solid nut embracing the screw, and geared to a handle on the right of the slide-rest, the rotation of which moves it along the bed. When, however, the handle is prevented from rotating by engaging a pawl with a toothed wheel attached to it, the nut is locked, with the result that the rotation of the screw gives motion to the saddle. A stop placed upon a rod running the whole length of the bed, and passing through the saddle will cause the saddle to disengage a clutch in the reversing gear and stop the traverse at a predetermined point.

Another feature is the compact form of back gear, giving increased range of speed, which was probably originated by Roberts. A countershaft is placed at the back of the headstock, and carries a spur wheel and pinion gearing with those on the pulley and mandrel; its bearings are supported in horizontal slides, which are actuated by two cam levers connected by a handle bar, so that it may be easily put into or out of gear.

17 Planing Machine

This machine, made by Richard Roberts in 1817, is believed to be the earliest planer now in existence and one of the first made for planing metal.

In general design the machine bears a strong resemblance to the modern planer. The work is bolted to a table which moves to and fro on a straight bed, under a fixed tool which is capable of being traversed so that, by the two motions, plane surfaces are produced as the tool makes successive cuts. The table is mounted on one side on a narrow flat face and on the other on an inverted V and is hand operated by means of a chain and drum. The cross slide is supported on two standards bolted to the bed and is provided with vertical adjustments by two screws which are not however interconnected. It has an internal screw for traversing the tool rest which is capable of angular movement and is fitted with a separate hand feed motion for the head. The tool is held in a hinged clamp which allows it to lift on the return stroke.

18 Slotting Machine

This machine was manufactured by Sharp, Roberts & Co. in about 1830 and was used for cutting the keyways in waggon wheels for the Manchester and Liverpool Railway.

The slotting bar is mounted vertically in a cast-iron frame and obtains a reciprocating motion from a connecting rod and crank wheel on the spindle which carries the flywheel and driving pulley. The crank pin is secured in a groove in the crank wheel and may be adjusted to give any stroke up to 6 in.

The table which holds the work may be rotated and may be moved horizontally in slides. The latter motion can be made self-acting. The table frame may also be adjusted vertically and slides between rectangular ways on the inside of the frame standards.

The introduction of a machine for key-grooving was attributed to Richard Roberts by both Nasmyth and Holtzapffel. It obviously derives from Brunel's mortising machine (13).

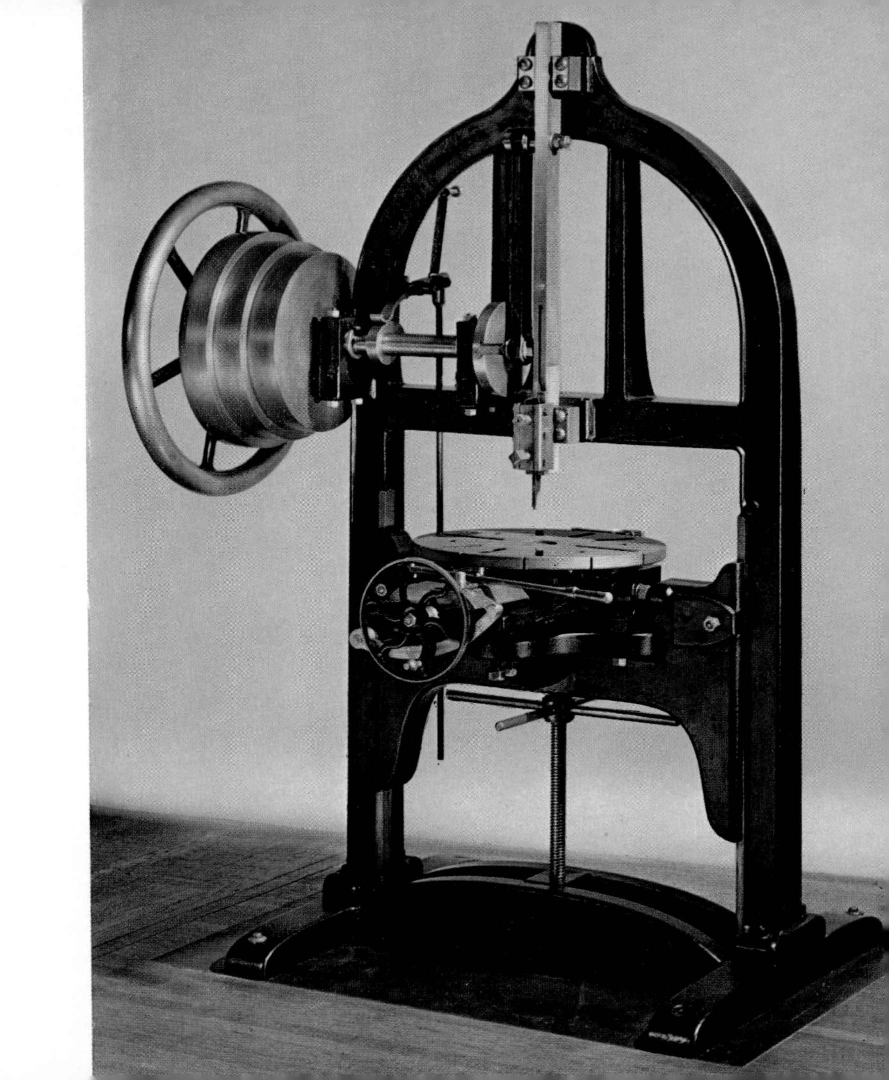

Science Museum illustrated booklets

Other titles in this series:

Aeronautica Objects d'Art, Prints, Air Mail

Aeronautics Part 1: Early Flying up to the Reims meeting

Aeronautics Part 2: Flying since 1913

Agriculture Hand tools to Mechanization

Astronomy Part 1: Globes, Orreries and other Models

Astronomy Part 2: Astronomical Telescopes

British Warships 1845–1945

Cameras Photographs and Accessories

Carriages to the end of the 19th century

Chemistry Part 1: Chemical Laboratories and Apparatus to 1850

Chemistry Part 2: Chemical Laboratories and Apparatus from 1850

Chemistry Part 3: Atoms, Elements and Molecules

Chemistry Part 4: Making Chemicals

Clocks and Watches Part 1: Weight-driven

Clocks and Watches Part 2: Spring-driven

Fire Engines and other fire-fighting appliances

Lighting Part 1: Early oil lamps, candles

Lighting Part 2: Gas, mineral oil, electricity
Lighting Part 3: Other than in the home
Making Fire Wood friction, tinder boxes, matches
Microscopes
Motor Cars up to 1930
Motorcycles
Physics for Princes The George III Collection
Power to Fly Aircraft Propulsion
Railways Part 1: To the end of the 19th century
Railways Part 2: The 20th century
Sewing Machines
Ship Models Part 1: From earliest times to 1700 AD
Ship Models Part 2: Sailing ships from 1700 AD
Ship Models Part 3: British Small Craft
Ship Models Part 4: Foreign Small Craft
Steamships Part 1: Merchant Ships to 1880
Steamships Part 2: Merchant Ships from 1880
Surveying Instruments and Methods
Textile Machinery
Timekeepers Clocks, watches, sundials, sand-glasses

Published by
Her Majesty's Stationery Office
and obtainable from the
Government Bookshops listed
on cover page iv (post orders
to PO Box 569 London SE1)

35p each (by post 40p)

Printed in England for
Her Majesty's Stationery Office
by Heffers Printers Limited
Cambridge

Dd 505767 K60